Discovery of Structure
&
Formation of Photons

Dr Abhijit Thakur

DEDICATION

Sir Isaac Newton by Sir Godfrey Kneller, 1689(PD-Art)

This book is dedicated to Sir Isaac Newton & Newtonian Physics, which can be extended to define Atomic and Subatomic Particle Interactions as well as Galactic structure, function and Cosmic events without Distorting and Twisting Nature

CONTENTS

ACKNOWLEDGMENTS

I acknowledge unconditional support of my parent, Dr Sudha Thakur, Dr Ramchandra Thakur and wife Dr Priya Thakur and very affectionate Arya and Anushka

1 STRUCTURE AND FORMATION OF PHOTON

Defining Elementary Particle or Fundamental Particle or God Particle or Brahma Particle is the easiest. It is the smallest, the fastest, superluminal particle with its physical properties are physical properties of matter stretched to perfection. It is the one and only elementary particle. (Thakur, A. (2016) Super Unified Theory)

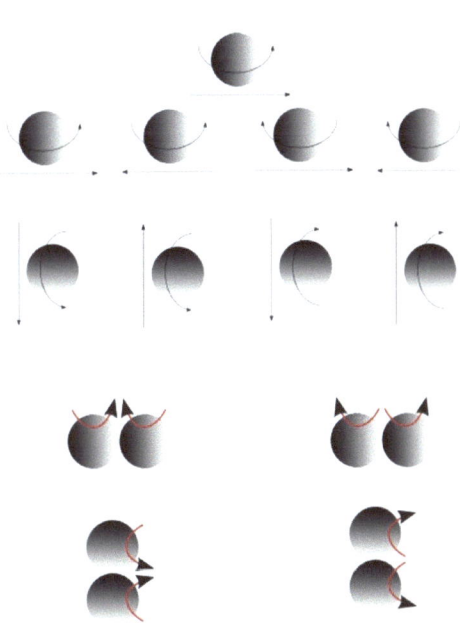

Figure: Brahma particle or God Particle is PERFECTLY SPHERICAL, PERFECTLY SOLID, PERFECTLY RIGID, NON-DEFORMABLE PARTICLE WITH MASS AND ENERGY. It is non-thermal, non-electric, non-magnetic, perfect perpetual machine with perfect conservation of mass and energy separately. It has linear and angular motion, the contact alignment of Brahma particle is defined by non-conflict counter spinning alignment.

Brahma Particle forms Universe by forming supporting walls of Universe and forming a Basic cosmic flow across the supporting walls of Universe. BRAHMA PARTICLE'S FREE FORM FLOWS IN BASIC COSMIC FLOW AND COMPOSITE FORM FORMS ALL THE MATTER FORMS OF UNIVERSE AND UNIVERSE SUPPORTING WALL.

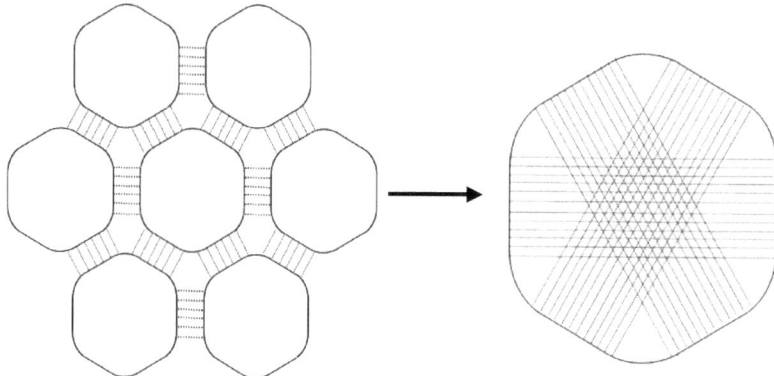

Figure: Brahma Particle or God Particle in "SPIN ALIGNED CLOSE PACKING" form a Universe polyhedron, which has multiple flat faces joined by smooth curvature without vertex. Universe is populated by Brahma particles as basic cosmic flow and composite forms of Brahma particle.

Brahma particles close packing is SPIN ALIGNED CLOSE PACKING

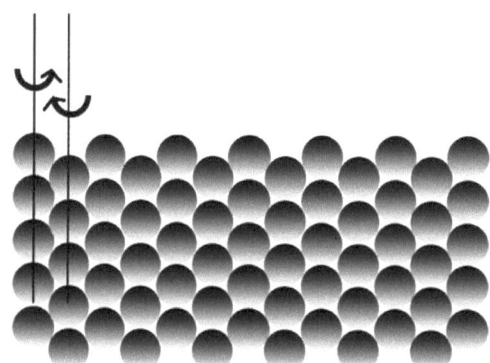

Figure: In one plane counter-spinning columns of Brahma Particle forming close packing arrangement, "SPIN ALIGNED CLOSE PACKING".

"SPIN ALIGNED CLOSE PACKING" is soul of Brahma Particle's formations and in all its composite forms whether it's supporting wall of universe or SHELL OF PHOTON BODY, this close packing is represented.

Figure: Table top model of spin aligned close packing, spinning central row is in non-conflict spin contact at node with four (4) particles right angle to each other. The pattern forms hexagonal close packing in vertical and horizontal plane while in coronal pane its body centred close packing, termed as **INTERLOCKED SPIN ALIGNED CLOSE PACKING.**

Depicting similar pattern with spin directions, finally this spin aligned close packing will be seen in shells of Photons.

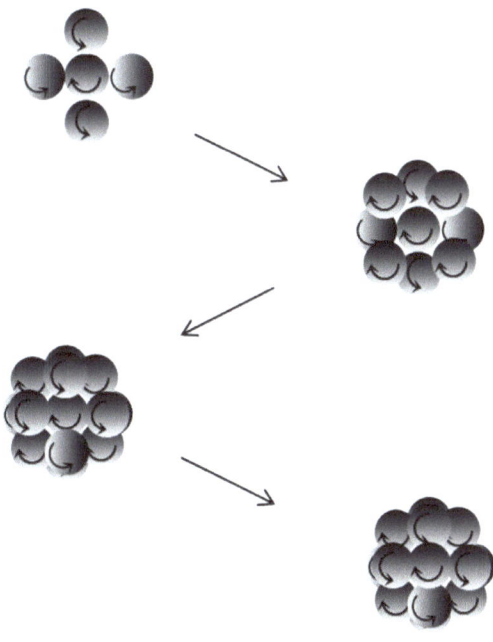

Let's see how Brahma Particle forms supporting wall of universe

 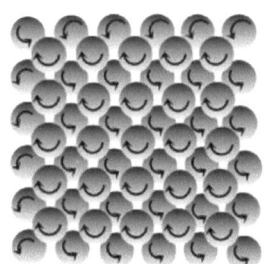 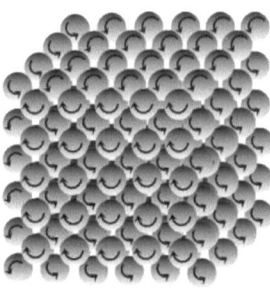

Figure: Brahma Particle with spin along the plane of supporting wall in spin aligned close packing forms flat faces of universe polyhedron joined by smooth curvature. Exposed surface has array of depicted surface units, forming a mosaic of convex reflecting surfaces. This reflecting surface has Randomizing effect on any impacting Brahma Particle flow patterns.

Brahma Particle forms basic cosmic flow across the walls of Universe. Basic cosmic flow is multidirectional, bilaterally symmetrical, parallel, uniform and steady flow in the space of Universe.

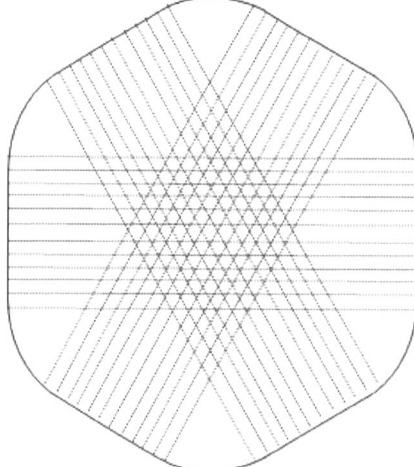

Figure: Multiple faces (12, 14, 16, 18 or…, debatable) of Universe have Randomizer effect on reflection. Final result is multidirectional, bilaterally symmetrical, parallel, uniform and steady flow across walls of Universe, The basic cosmic flow of Brahma Particle. There are 6, 7, 8, 9 or…, debatable, major bilaterally symmetrical beams, incorporate basic cosmic flow in the space of universe.

Free form of Brahma particle flows in Basic cosmic flow and composite form represent all the Matter forms in Universe. Composite form is unpacked in core of Black Holes and Free Form is released. Progressive rise in free form brings about congestion, turbulence in Basic cosmic flow, initiating a cascade of events leading to Catastrophic dissolution of Universe. Catastrophic dissolution adds up huge load of free form of Brahma Particle to Basic cosmic flow. This creates multiple areas of congestion in basic cosmic flow. Periphery of these areas of congestion becomes zone of formation of vortices. The stable vortices represent repopulation of universe with composite forms of Brahma particle again, what vanished in catastrophic dissolution. The specialized stable vortices are future PHOTONS of Universe.

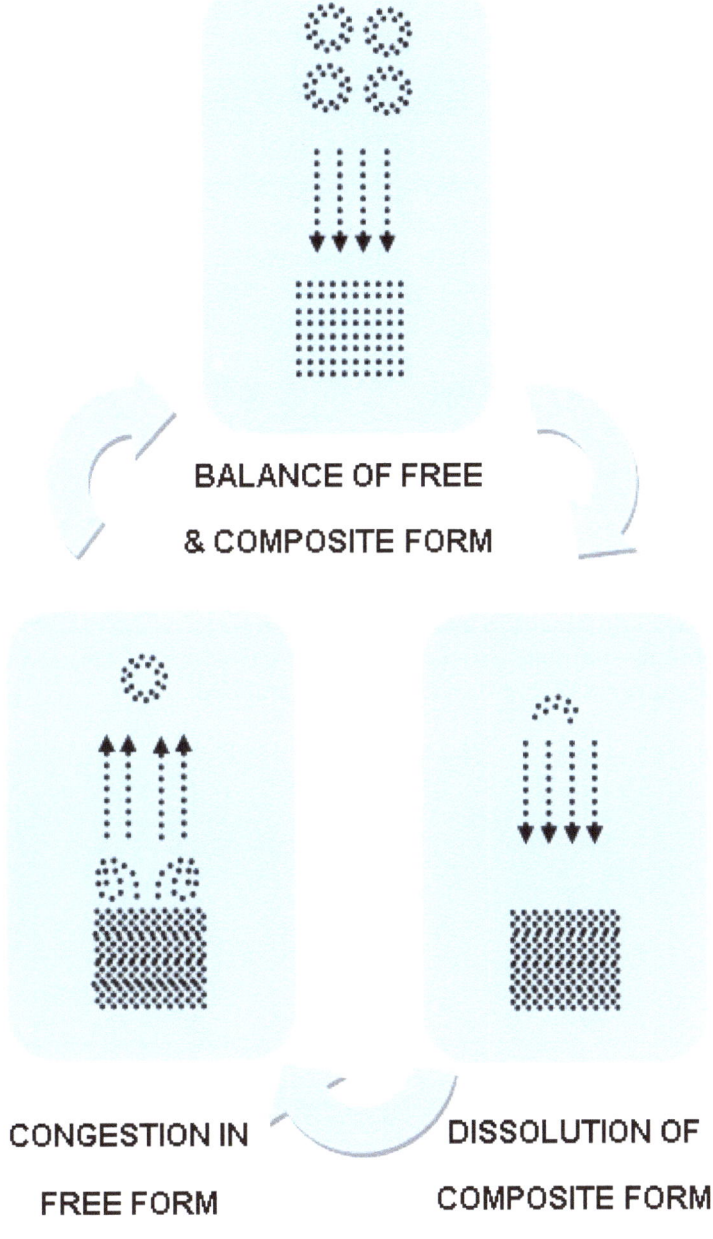

Basic cosmic flow density progressively rises along cosmic cycle reaching critical value where its flow starts congesting heralding catastrophic dissolution of Universe, post catastrophic dissolution as universe starts stabilizing a process of assembling of free form Brahma Particle into composite form is initiated. (Thakur, A. (2016) Super Unified Theory)

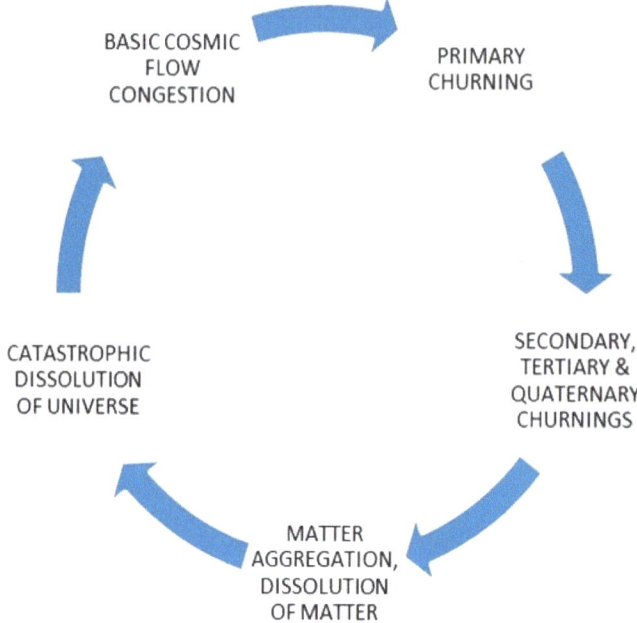

Figure: Cosmic Cycle, presently we are in phase of matter aggregation and dissolution of matter. Most of the phases of cosmic cycle are non-observable phases of cosmic cycle. Catastrophic dissolution is eminent in astronomical time frame (50-60 billion years). In the phases following Catastrophic dissolution, Universe is populated with its composite particle, which take almost 150 to 200 billion years till event of Big Bang is reached.

Basic cosmic flow has free flowing independent Brahma particles with Multidirectional components, bilaterally symmetrical, laminar, uniform and steady flow. On congestion Brahma particle starts forming simple composite forms by forming spin aligned close packing. This spin aligned close packing is the simplest composite form of Brahma Particle. SPIN ALIGNED CLOSE PACKING initiates with pleats to beams of CUBIC CLOSE PACKING in spin alignment to INTERLOCKED SPIN ALIGNED CLOSE PACKING. The complex composite forms of Brahma Particles with functionality would be astonishing

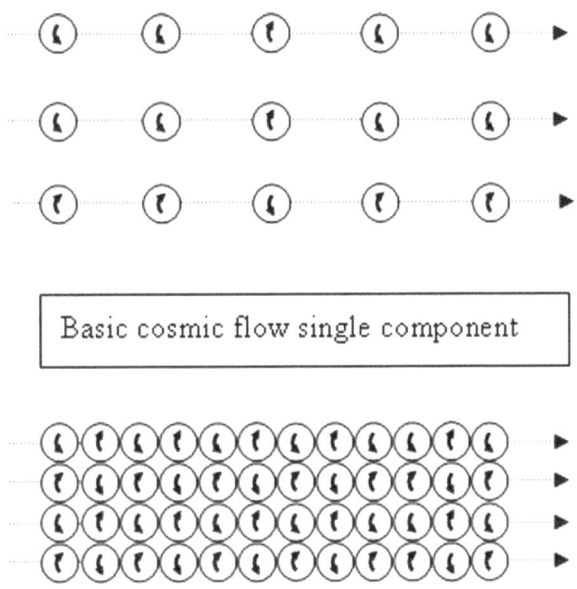

Basic cosmic flow single component

Alignement and adjacent counter spin on

controlled compression

Figure: One of the simplest Composite form of Brahma Particle, SPIN ALIGNED Cubic close packing. On further compression it would progress to INTERLOCKED SPIN ALIGNED CLOSE PACKING. Towards zone of congestion contact counter-spinning or non-conflict contact spinning arrays start taking shapes. *"The complex composite shapes revealed by Nature through such process have amazing functionalities"*.

Around the zone of stagnation, ANALOGOUS TO KARMAN VORTICES, vortices of beams of close packed brahma particle start taking shape.

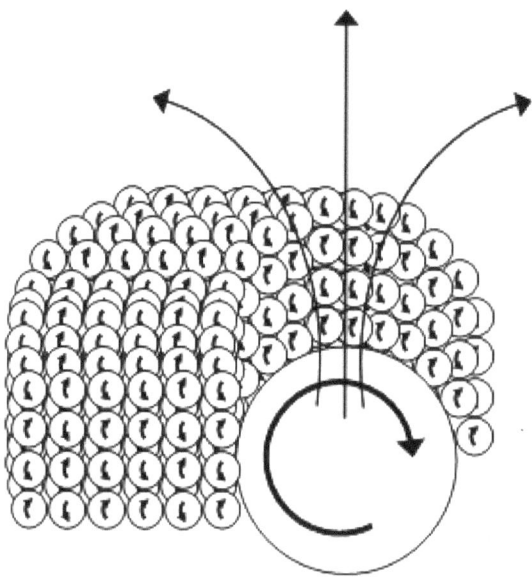

Brahma Particle close packing entering into curved formation with internal correction by ejection or displacement of adjacent particles. This is the beginning of process of vortices formation around zone of stagnation; the stable vortices represent future photons.

The process of formation of vortices around a zone of stagnation is termed as Primary Churning. Though the process is not a typical churning process described in Secondary Churning, Tertiary Churning and Quaternary Churning but as continuity it is termed as Primary Churning. Secondary Churning is processes of formation of electron, positron and Neutrino, Tertiary Churning is process of formation of proton and Neutron, and Quaternary Churning is process of formation of atomic nuclei larger than hydrogen.

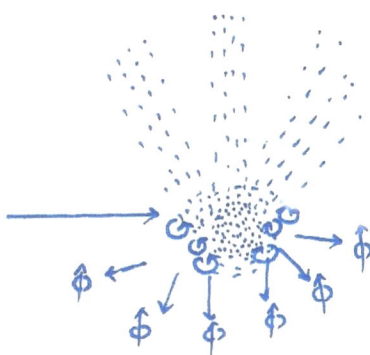

Figure: At the periphery of zone of congestion vortices are formed and released into space in pattern similar to Karman vortices with difference that vortices are formed in all major axes.

A pictorial depiction of Primary Churning, the process of Formation of Photons, Photons are first generation composite particles of Brahma Particle.

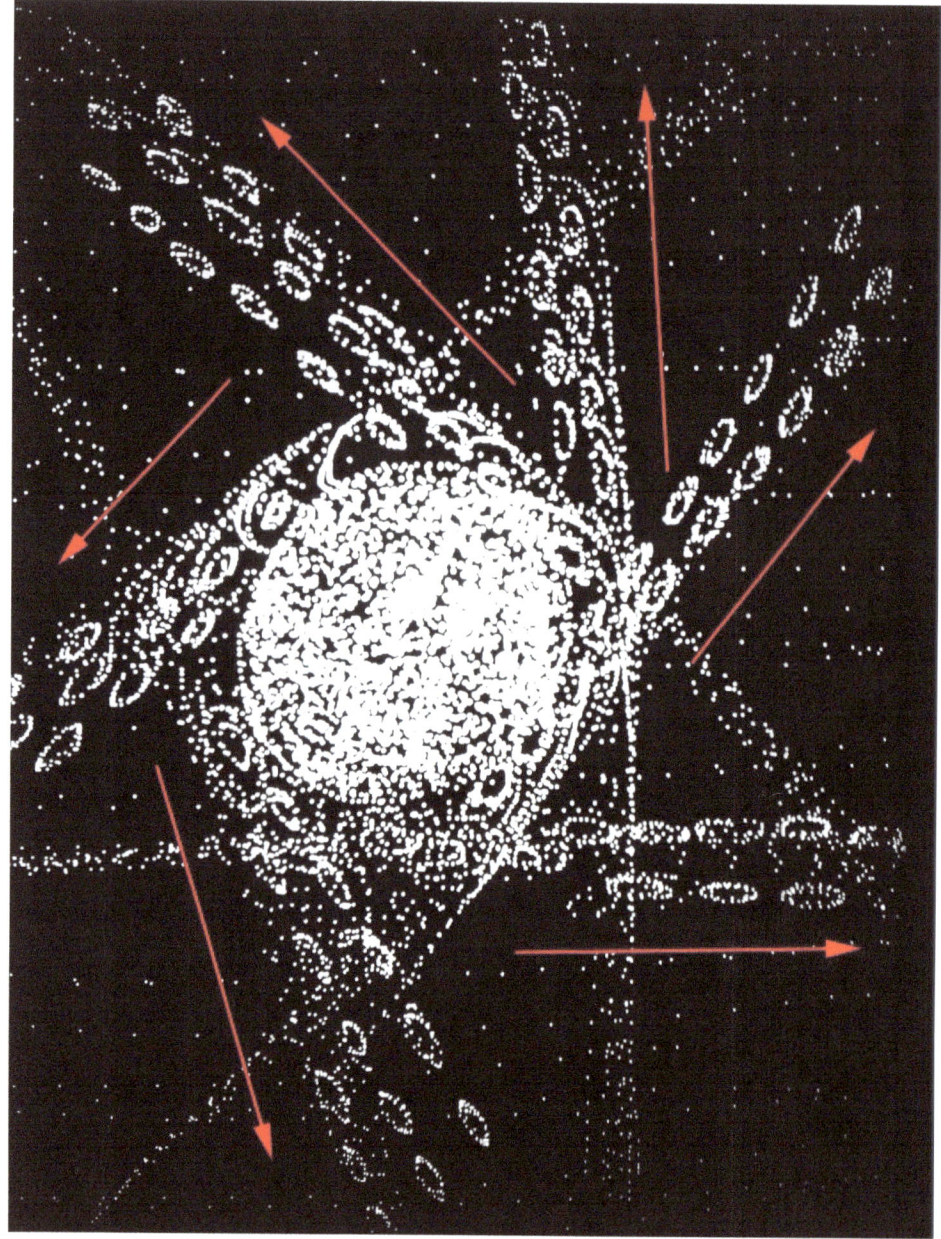

Figure: Areas of stagnation are formed in the central zone of congestion, surrounding this zone of congestion there are formation of vortices. These vortices are released into the space of Universe around the zone of congestion. The stabilized vortices are future photons, but journey to form EMR is still very long.

The released vortices are finely carved in the space of basic cosmic flow. Some special characteristics are needed for their survival in the space of Universe. The special process of standardization is also designed by Universe for these photons. Thakur, A. (2016) Super Unified Theory

A specialized vortex has counter spinning row of Brahma Particle along its prolate spheroid body with spin axis of spinning rows along equator and latitudes. This prolate spheroid has Polar Defects with Polar Gating. This specialized prolate spheroid is future Photon.

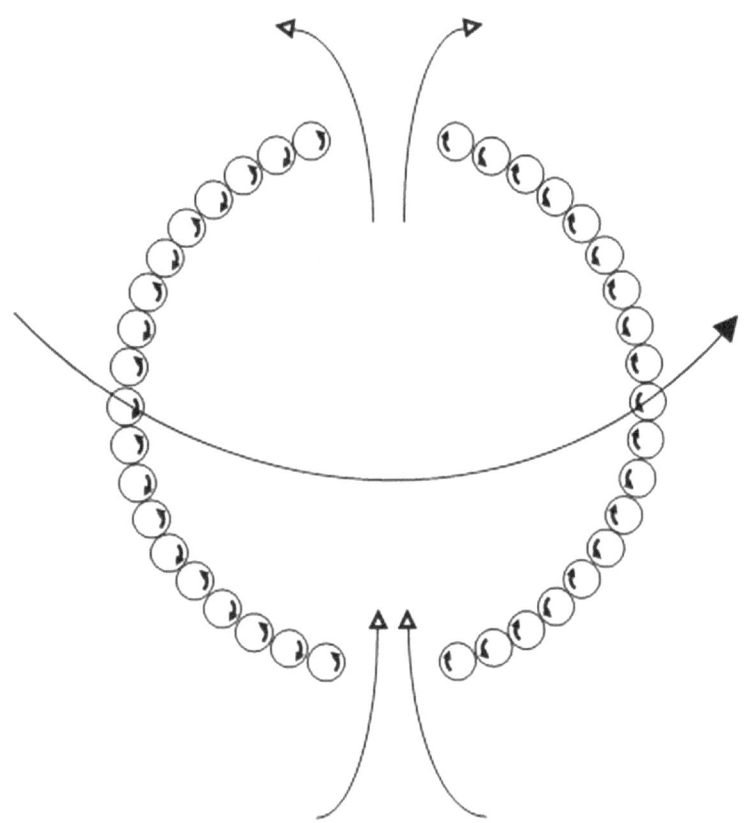

Figure: **Innermost layer of shell of Photon, with counter-spinning rows, Polar Gating is achieved by the spin of the row lining Pole. Inflow pole represent South Pole and Outflow Pole represents North Pole of PHOTON.**

INTERLOCKED SPIN ALIGNED CLOSE PACKING is soul of composite formations of Brahma Particle, Various aspects of finer correction in structure of Photon is achieved in the environment of Basic cosmic flow. Thickness of shell of Photon is subject to interactions of multiple forces acting on body of Photon in environment of Basic cosmic flow.

"DON'T BE SURPRIZED GRAVITY AND GRAVITATIONAL FIELDS HAVE REAPPEARED IN UNIVERSE AFTER TOTAL ELIMINATION DURING CATASTROPHIC DISSOLUTION. THIS REAPPEARANCE IS IN FORM OF GRAVITY AND GRAVITATIONAL FIELDS OF PHOTONS".

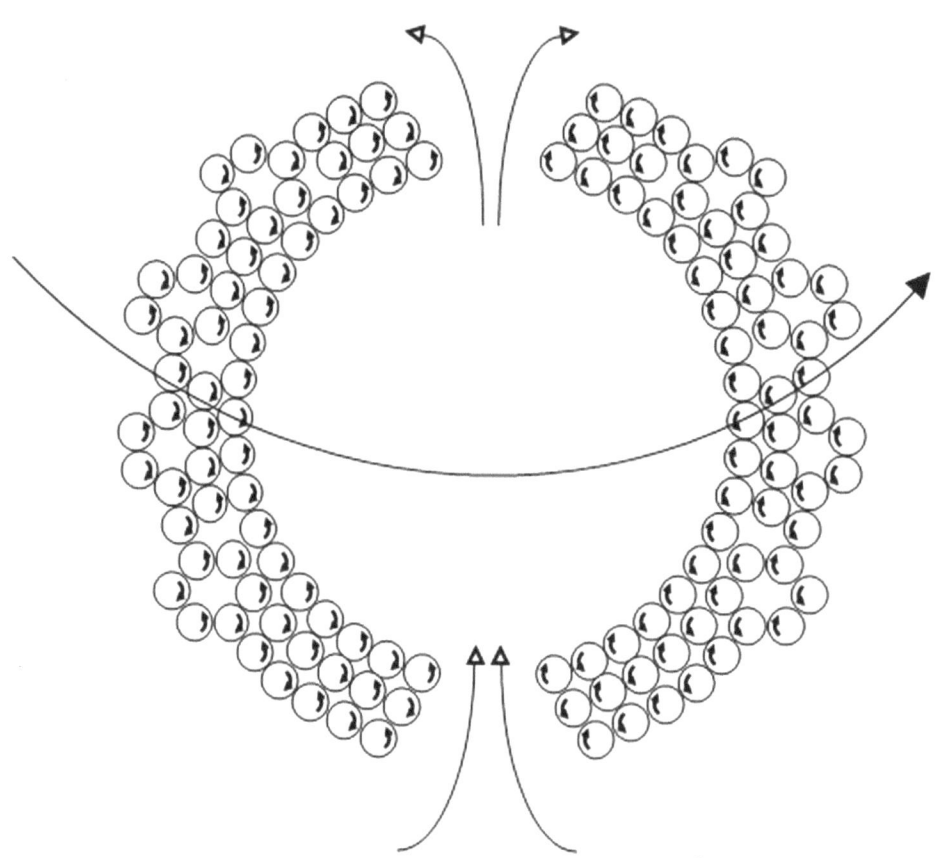

Figure: The shell of Photon with additional layers of rows of Brahma particle. The basic principle of this formation that is spin aligned close packing is soul of this formation. For depiction purpose spin aligned cubic close packing is shown, in true form interlocked spin aligned close packing is expected.

INTERLOCKED SPIN ALIGNED CLOSE PACKING has got transformed from simple composite form to complex composite form. This complex composite form in shape of PROLATE SPHEROID WITH POLAR DEFECTS is gradually transformed into a fully functional PHOTON.

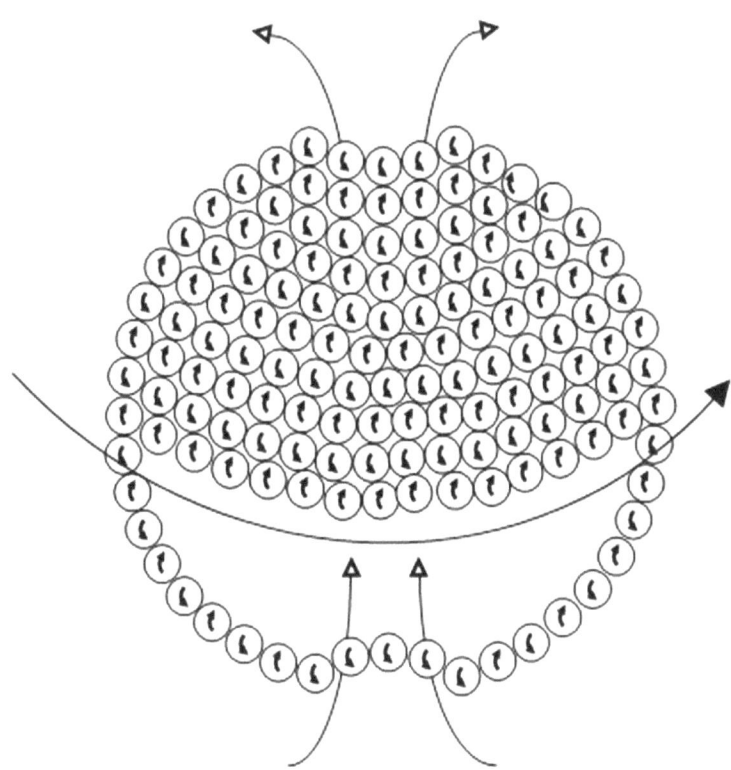

Figure: Counter spinning adjacent rows of Brahma particle, following principle of spin aligned close packing, form layers of shell of PHOTON. Additional layers add up following interlocked spin aligned close packing with internal adjustment by ejection or displacement of Brahma particle under constrain. Polar gating defines inflow South Pole and outflow North Pole. The body of PHOTON acquires spin along polar axis with forward movement.

Shell of photon is stabilized into the space of basic cosmic flow by compression force of basic cosmic flow equivalent to nuclear force counterbalanced by centrifugal force due to spin of photon along its polar axis. The stabilization is finally accomplished under a buffer of layered MICROENVIRONMENT of PHOTON described ahead.

Table-top model view of innermost layer of photon from inside **PHOTON** depicting its **SPIN ALIGNED CLOSE PACKING**

Figure: Innermost layer of shell of PHOTON, they are in interlocked spin aligned close packing

Figure: Constrains of curvature are managed by ejection or displacement of adjacent brahma particle maintaining spin aligned close packing

The shell of PHOTON into the space of Basic Cosmic flow acquires MICROENVIRONMENT of free form of Brahma Particle by gravitational interaction. Microenvironment is complex interplay of gravity, spin, speed, polar gating and shape of the particle concerned. (Thakur, A. (2016) Super Unified Theory)

PHOTON shell is spinning along its polar axis and has a speed perpendicular to polar axis.

IF PHOTON MOVES WITH SPEED OF LIGHT, BRAHMA PARTICLE HAS TWO ADDITIONAL DIMENSION OF MOTION.THESE DIMENSIONS IN ADDITION TO SPEED OF PHOTON, ARE SPIN ALONG THE SHELL OF PHOTON AND SPIN OF BRAHMA PARTICLE ITSELF, THIS IMPLIES BRAHMA PARTICLE IS SUPERLUMINAL.

Photon develops three distinct MICROENVIRONMENTS
1. Orbital Microenvironment
2. Polar Microenvironment.
3. Electric Microenvironment
4. Internal Microenvironment

Orbital Microenvironment is basic microenvironment of a composite particle, covering the shell of the particle. It creates a buffer zone of layers of curvilinear trajectories of brahma particles, curving along the spin of the particle. Orbital microenvironment follows the spin of the particle. It propagates the physical properties of the particle and carries along the spin of the particle.

Dilution of perfect physical properties of brahma particle is achieved through buffering along the thickness of orbital microenvironment. Physical interaction in form of collision of particles, friction, and deformability is mediated through orbital microenvironment. Perfect physical properties of brahma particle are diluted to observed physical properties of matter due to buffering across orbital microenvironment. These interactions between two bodies could be explained based upon interaction amongst orbital microenvironment of enumerable participating particles in matter forms

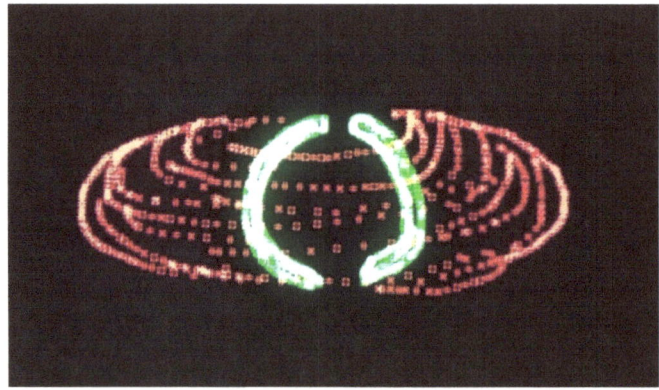

Figure: Free form of Brahma particle with progressively changing property status from high spin low speed to low spin high speed form a cover around shell of PHOTON body like an environment around a planetary body. This MICROENVIRONMENT is formed by curvilinear trajectory of Brahma particle with specific property status, around shell of PHOTON by gravitational interaction. (Thakur, A. (2016) Super Unified Theory)

Polar Microenvironment

Polar Microenvironment is generated around PHOTON due to its polar gating. It is generated by layer of trajectories of brahma particle radiating from outflow pole or North Pole. These trajectories propagate as Polar Jets or Cylindrical Columns and curvilinear trajectory across the body of the particle reaching the inflow pole or South Pole of the particle. It is responsible for magnetic properties of the particle. Polar microenvironment created by brahma particle flow pathways, on complex mixing represents magnetic field lines. Magnetic interaction between two bodies could be defined on the basic of these patterns of brahma particle flow and resultant mechanical interaction at the level of poles of individual particle...

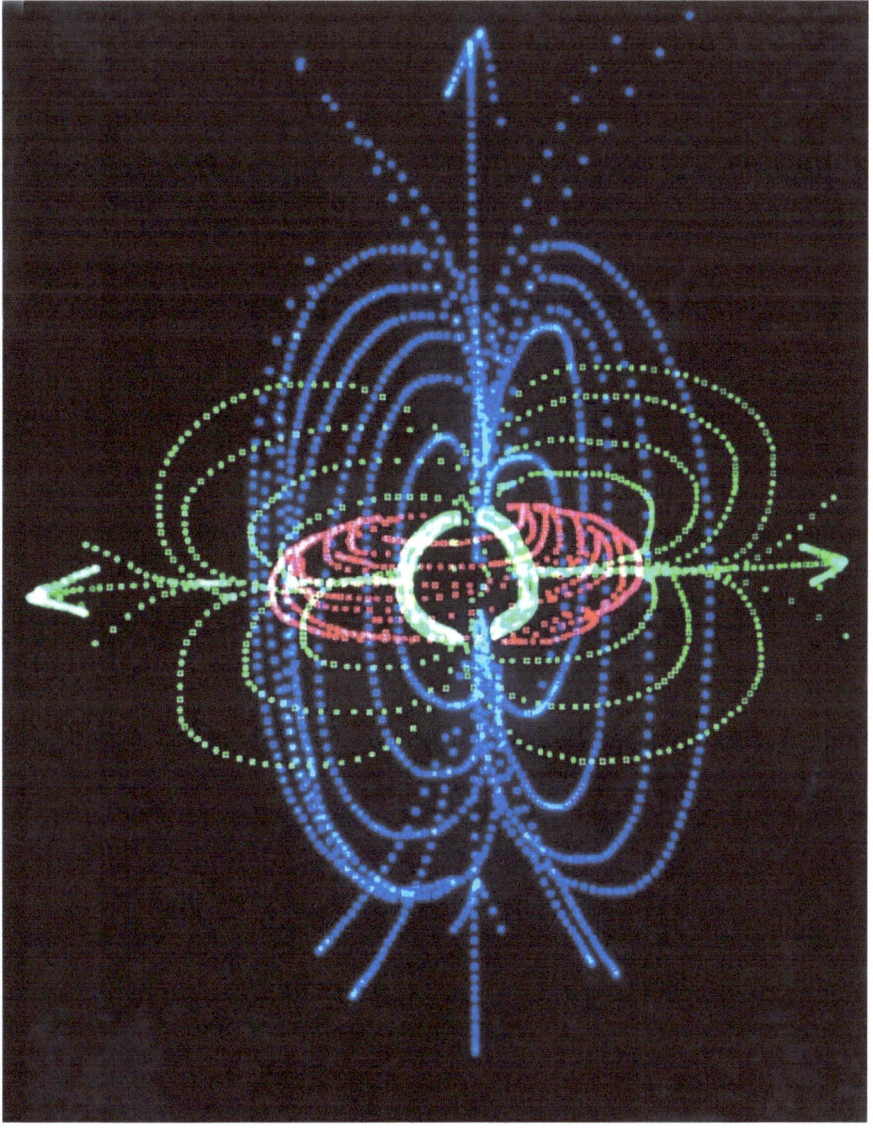

Figure: Blue lines represent selected trajectories of Brahma particle, formed due to Polar gating in PHOTON, making its polar microenvironment.

ELECTRIC MICROENVIRONMENT is a specialized microenvironment, it is more elaborate in Electron, Positron and Proton but its primitive form it is represent in PHOTON as well.
Electric or Equatorial microenvironment
Equatorial microenvironment is a specialised microenvironment of PHOTON, which holds electric properties. It originates along equatorial plane of the particle and extends towards poles. Equatorial microenvironment imparts electronegativity to PHOTON. (Thakur, A. (2016) Super Unified Theory)

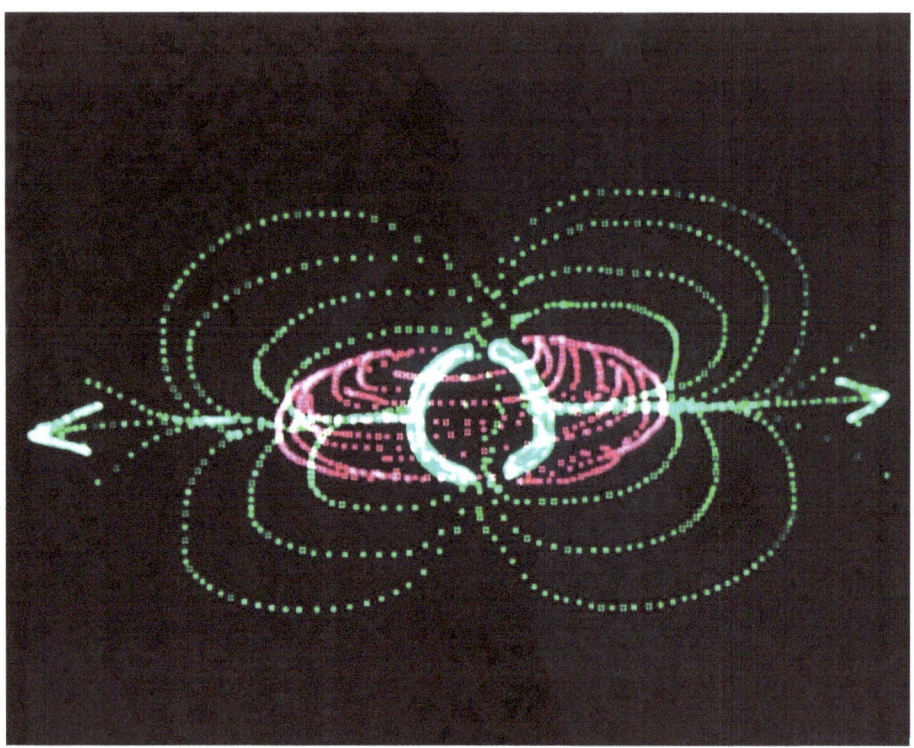

Figure: Curved expose shell of generates one more specialized flow pattern of Brahma particle around PHOTON called Electric Microenvironment. It propagates from equator of the particle and extend along latitudes to merge with polar flow. Its plane of action is equatorial plane corresponding to electronegativity of PHOTON

INTERNAL MICROENVIRONMENT is layered microenvironment of free form of Brahma particle present inside the shell of photon. It partially supports structure of shell from inside.

Computer generated Model of PHOTON; all images are not to measure

Prolate spheroid shaped Photon shell formed by multi-layered array of brahma particle in interlocked spin aligned close packing with polar defects. The rows of Brahma particle have spin axis along equator and latitudes of photon shell.

On release of PHOTON in the space of basic cosmic flow it acquires buffering, protective and functional covering of free form of Brahma particle, computer generated model, not to measure

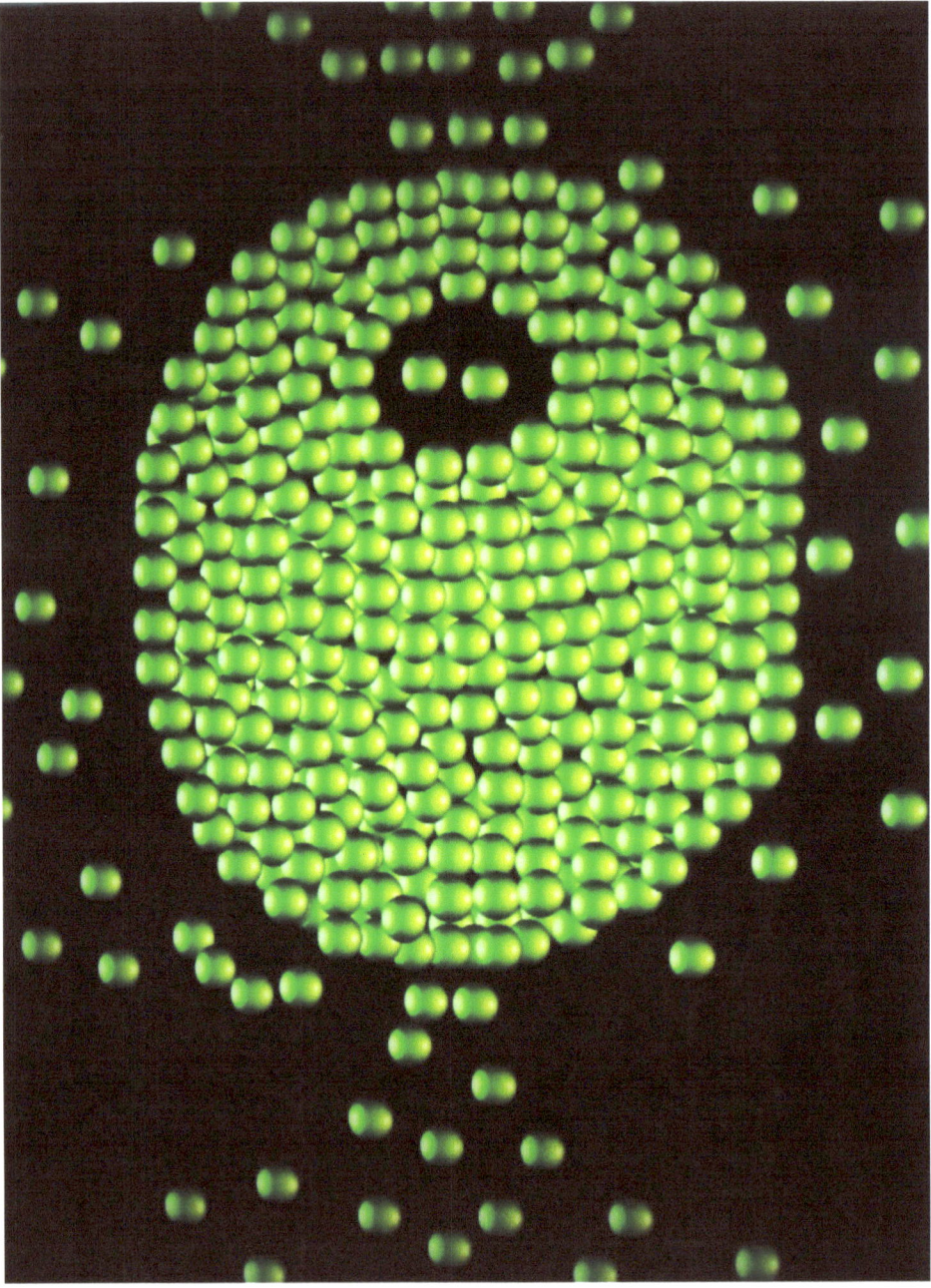

Figure: Just for depiction photon shell with orbital and polar microenvironment, microenvironment is very dense and rarefies with distance. Very thin microenvironment is shown for display purpose, not to measure.

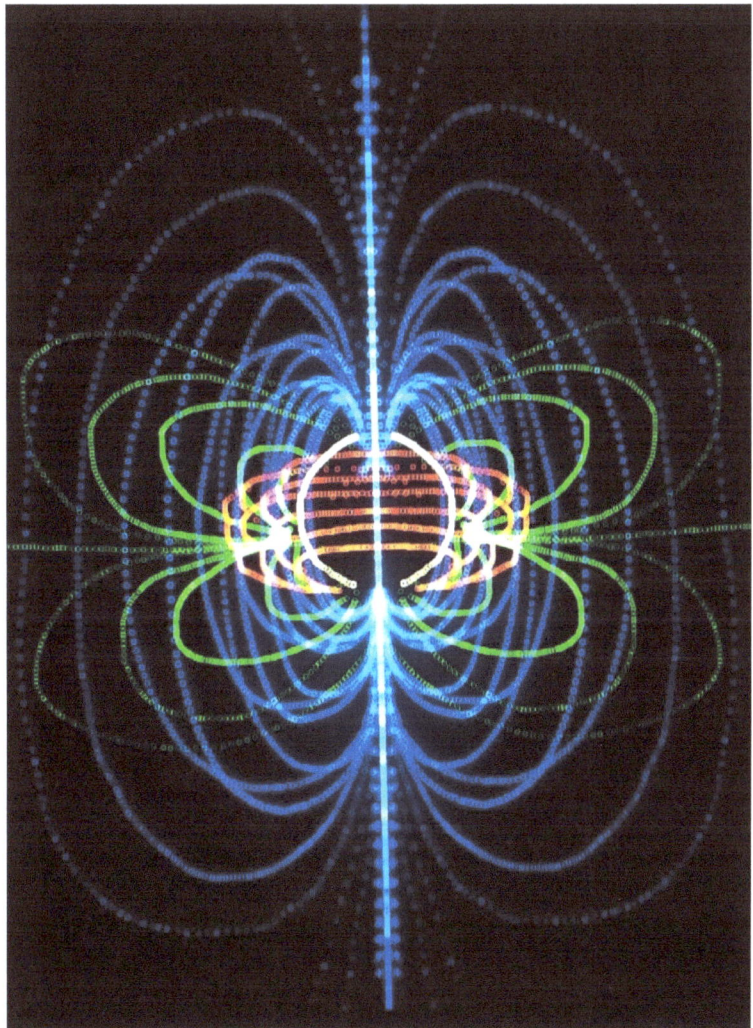

Figure: Central shell has covering of orbital microenvironment depicted in PINK, specialized electric microenvironment due to shape, depicted in GREEN and specialized polar microenvironment formed due to Polar Gating achieved by spin of brahma particle lining poles depicted by BLUE.

The fastest smallest stable photons win the race. Large, sluggish, unstable, defective PHOTONS meet their fate by colliding with smaller and faster photons showering from their site of origin. Smaller, faster and stable photons rip through larger ones along their journey form their site of origin at periphery of Brahma particle congestion zones towards limiting wall of Universe. (Thakur, A. (2016) Super Unified Theory)

Photons have long journey to follow. IT MAY BE AROUND 30 TO 40 BILLION LIGHT YEARS BEFORE THEY GET CONVERTED TO ELECTROMAGNETIC RADIATIONS. This is the journey of photon from their site of origin to limiting wall of universe and then focussing back to the reciprocal points in the space of universe. Important point to notice that supporting wall of universe has no randomizer effect on photon reflection because its surface is also mosaic of exposed bodies of Brahma Particle.

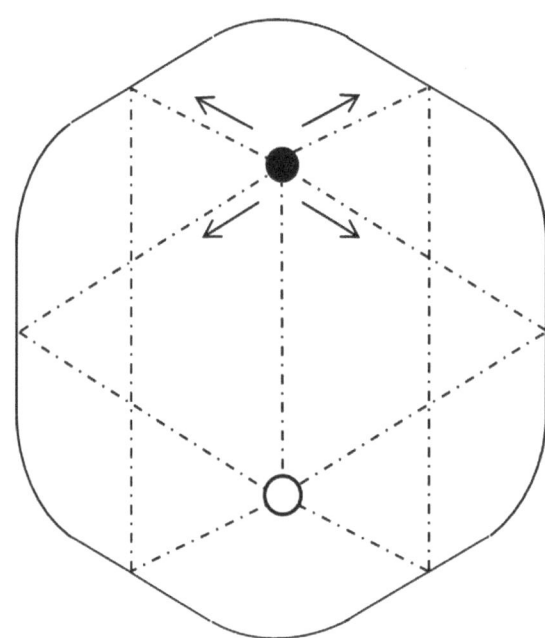

Figure: Starting from periphery of Zone of congestion to limiting walls of Universe and then focussing back to reciprocal points complete photon trail. This is trail is repeated many times and has performed major functions of forming ELECTRON, POSITRON & NEUTRINOS in SECONDARY CHURNING PROCESS. The weakened trail is imprinted in present Universe as CMB.

PHOTON TRAIL becomes very important when it starts focussing at reciprocal point in space of Universe. This initiates two (2) very important processes, firstly perfect standardization and alignment of photons into EMR and secondly, Formation of swirling ball at focussing points called, secondary churning balls are site of formations electron, positron and neutrinos.

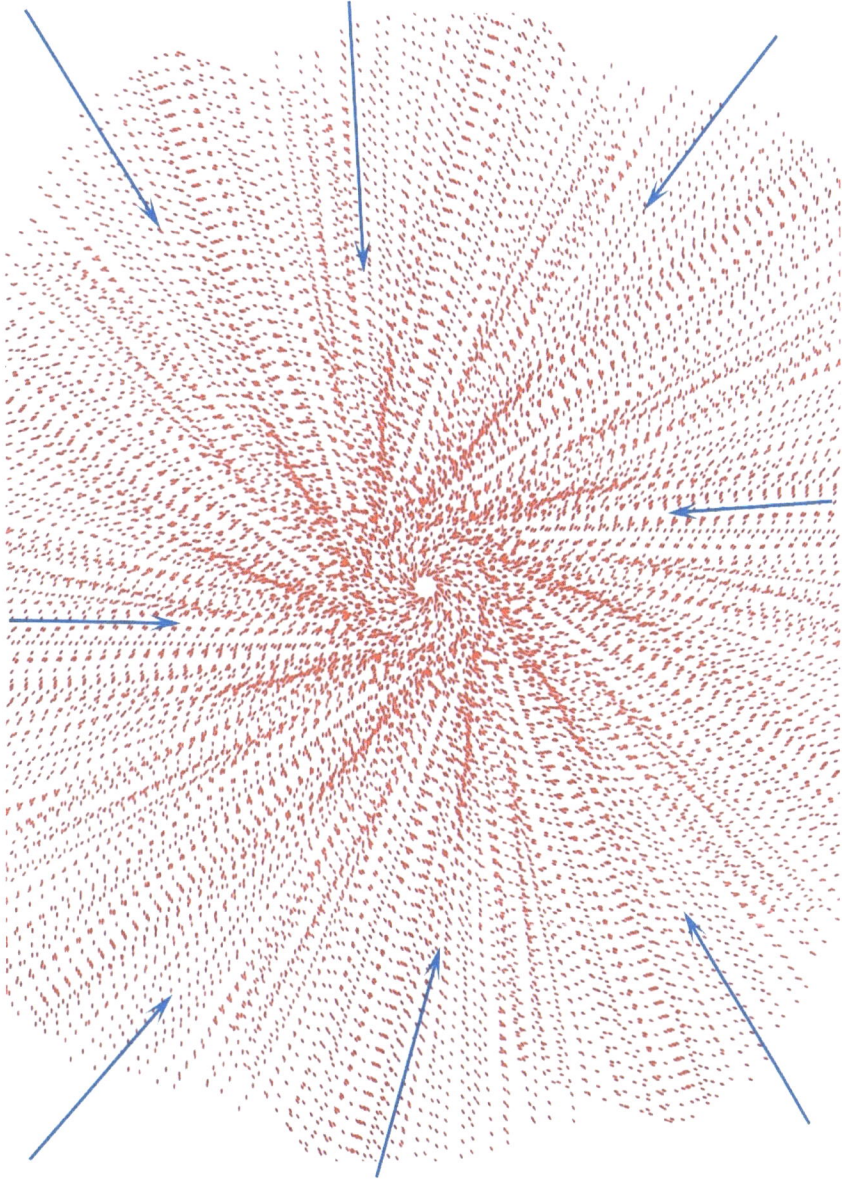

Figure Enumerable, countless photons following the photon trail focus back to reciprocal points as billions of light yearlong beams of Gamma ray LASERS to energies secondary churning, releasing at periphery EMR as by-product to represent CMB in today's universe after multiple trips to limiting wall of Universe. Secondary churning is subject of next release in series of pictorial summary of Super Unified Theory. (Thakur, A. (2016) Super Unified Theory)

Focusing beams of Photons brings about perfect standardization of photon. Photon with perfectly matched spin and speed could remain in the track, while mismatched photons are ejected or left behind to be destroyed. Linear compression in such beam leads to alignment of photons with their polarity and spin. ALL PHOTONS WITH PERFECTLY MATCHED SHELLS ARE INCORPORATED IN SUCH FORMATIONS CALLED EMR; VARIABLE ENERGIES ARE SUBJECT TO VARIABLE THICKNESS OF INTERVENING MICROENVIRONMENT OF PHOTONS IN FORMATIONS.

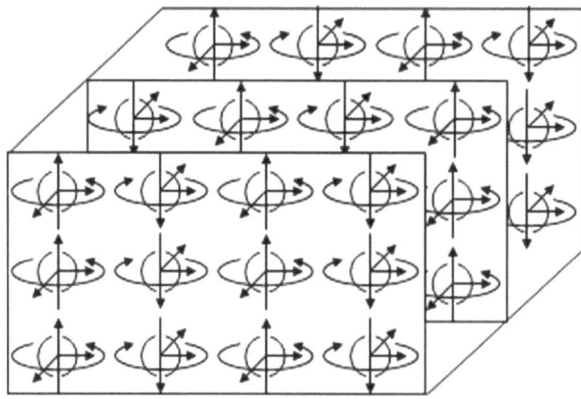

Figure: Beams of EMR with alternating polarity in X axis and alternating electric activity along equatorial plane in Y axis. Polar, electric Microenvironments in the array gives wave characteristic to EMR, Particle characteristic would be mediated through orbital microenvironment.

EMR has speed of light in basic cosmic flow but in other medium or matter forms speed is dropped due to additional drag imparted by microenvironment coverings of such forms.

EMR has exactly similar shells of Photon, so that a perfect match of speed and spin could be obtained. Variations in orbital microenvironment thickness are responsible for variation in energies. Photons are aligned by spin and polarity and linearly compressed by specialized process in Nature.

High frequency EMR or high linear density

Medium frequency EMR or medium linear density

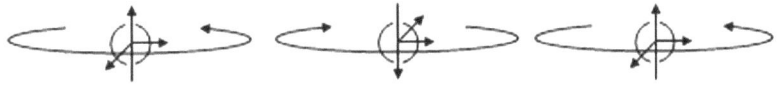

Low frequency EMR or low linear density

2 CONCLUSION

Structure and formation of Photon does not involve complex assumptions and difficult interpretations. It is a part of cascade of findings and discoveries revealed by Super Unified Theory. I cherish the treasure of Super unified Theory because it has answer for everything you ask. (Thakur, A. (2016) Super Unified Theory)

Super Unified theory has given me answers for not only structure and formation of photon but also that of electron, positron, neutrino, proton, neutron and even atomic nucleus. It has given answers for formation of galaxies, black holes, stars and their functionalities. Above all it has unified all fundamental forces of Nature.

There is no human manipulation in these findings and discoveries, as I hold hand of Brahma Particle; it takes me along its complete journey of Cosmic Cycle spread across 250 to 300 billion years. This journey is as amazing, as interesting, as mind-blowing and as exciting as nothing else.

.

ABOUT THE AUTHOR

Dr Abhijit Thakur is a Medical graduate with post graduate Medical degree in General Surgery. He is trained in advanced Laparoscopic surgeries. He holds Fellowship in Endocrine Surgery form La-Timone Hospital, Marseille, France. He is trained in surgical oncology in Mumbai, India. He is in active practise of Laparoscopic surgery and surgical oncology for last 15 years in South Mumbai.

He has many publications in Indexed Medical Journals and a book to his credit. His articles are common appearance in local Newspapers and magazines.

His grandfather Late Anant rao Yelnurkar was a highly qualified and skilled Civil Engineer. His legacy, directions and teaching were needed to be given a shape. Dr Thakur kept working off duty hours to mould the matter into some shape for last 25 years. With his mother Dr Sudha Thakur's pursuance, finally he took sabbatical form work and decided to complete the work in form of a book called Super Unified Theory

He had the belief that unbelievable will happen and to surprise one after other all the secrets of Universe were disclosed like a chain reaction. And ultimately the book Super Unified Theory was completed. When he read it back he himself could not believe that it has really happened!

Presently he lives in Mumbai, India.

Twitter.com/superunitheory

http://Superunifiedtheory.wordpress.com

http://Superunifiedphysics.wordpress.com

http://www.Superunifiedtheory.blogspot.com

Superunifiedtheory@gmail.com

https://www.amazon.com/author/abhijitthakur

www.ingramcontent.com/pod-product-compliance
Lightning Source LLC
Chambersburg PA
CBHW050407180526
45159CB00005B/2180